I0424403

Semblance of activity at the shared postsynapses and extracellular matrices:
A structure-function hypothesis of memory

Semblance of activity at the shared postsynapses and extracellular matrices:
A structure-function hypothesis of memory

Kunjumon I. Vadakkan

M.B.B.S, M.D (Biochemistry)
University of Calicut, India
M.Sc, Ph.D (Neuroscience)
University of Toronto, Canada

iUniverse, Inc.
New York Lincoln Shanghai

Semblance of activity at the shared postsynapses and extracellular matrices
A structure-function hypothesis of memory

Copyright © 2007 by Kunjumon I. Vadakkan

All rights reserved. No part of this book may be used or reproduced by any means, graphic, electronic, or mechanical, including photocopying, recording, taping or by any information storage retrieval system without the written permission of the publisher except in the case of brief quotations embodied in critical articles and reviews.

iUniverse books may be ordered through booksellers or by contacting:

iUniverse
2021 Pine Lake Road, Suite 100
Lincoln, NE 68512
www.iuniverse.com
1-800-Authors (1-800-288-4677)

Because of the dynamic nature of the Internet, any Web addresses or links contained in this book may have changed since publication and may no longer be valid.

The views expressed in this work are solely those of the author and do not necessarily reflect the views of the publisher, and the publisher hereby disclaims any responsibility for them.

ISBN: 978-0-595-47002-0 (pbk)
ISBN: 978-0-595-91286-5 (ebk)

Printed in the United States of America

To

All my teachers

Contents

Acknowledgements

I am indebted to the support from Neurosearch Centre, Toronto.

I thank Dr. Dagmar Novak for reading the manuscript.

Preface

It is a pleasure to present this short book to the readers. Here, we present *semblance hypothesis* that can explain memory and other higher brain functions. The hypothesis was evolved from a definition of memory based on synaptic function. Synaptic organization and patterns that can support this function were then proposed and partial evidence for these structural patterns is presented from the literature. The present hypothesis may become useful while searching for the engram.

Kunjumon I. Vadakkan
September, 2007

Summary

We suggest a hypothesis for memory based on a definition in terms of synaptic function. *A unit of memory, in the presence of an internal or external cue stimulus, results from the ability to induce specific postsynaptic events at the synapses of neurons from the learned item without the requirement of action potential (AP) reaching their presynaptic sides and which imparts an equivalent effect of AP reaching their presynaptic sides.* Organization of cellular and extracellular elements required is derived and structural evidence from the literature is presented. A highly plastic *shared postsynapse*[1] consisting of a cluster of postsynaptic membrane segments (PSMSs) each receiving separate presynaptic inputs is hypothesized. To begin with, excitatory postsynaptic potential does not spread from one postsynaptic segment to the next. Coincidence of APs reaching two axonal (presynaptic) terminals **A** and **C** of neurons **N1** (cue) and **N2** (item to be memorized) respectively, results in instantaneous membrane changes in the intervening area between their corresponding PSMSs **B** and **D** which functionally connect them. Preformed proteins are sufficient to meet these requirements. Repetition of learning induces multidirectional drifts in the positions of synapses on the shared postsynapses, bringing distally located co-activated PSMSs closer. These changes require new protein synthesis. During cue-induced memory retrieval, an AP reaching **A** depolarizes PSMS **B** and spreads to **D** possibly as depolarizing graded membrane potentials or induces postsynaptic cellular events on **D**. This generates the equivalent effect of activating presynapse **C** and evokes a cellular illusion of an AP-

---

1.   my italics

induced synaptic transmission from **C**. Similarly, AP which reaches one of the synapses among groups of single synapses on single dendritic spines separated by narrow extracellular matrix (ECM), named as *exneuron*[2], induces depolarization or cellular events in a second postsynapse by ionic effects through the ECM. The net effect of cellular illusions occurring in shared postsynapses and independent synapses through ECM effects in the network of neurons for the item to be retrieved results in *functional semblance*[3] for memory, a virtual sense of a stimulus in its absence. Each memory has a labile code consisting of the subset of postsynapses at the axonal terminals of neurons from the learned item that are activated during retrieval out of those that were co-activated with the cue during learning.

---

2.   my italics
3.   ibid

Introduction

The scientific nature of memory requires developing hypothetical frameworks that can be tested. Cajol's predictions [1] and Hebb's postulates [2] later found support from experimental long-term potentiation (LTP) [3]. The characteristic input specificity and associativity of LTP (review, [4]) were thoroughly explored for possible links with the mechanism of memory. Both spike-time dependent [5] and independent [6] synaptic plasticity changes were also examined for their physiological significance. In addition, dendritic spike-mediated co-operative LTP in selected clusters of synapses in dendritic domains [7] has been proposed to increase the memory capability of some neural networks (review, [8]). Behavioral studies also have shown an association between learning and LTP [9, 10]. Recently, a more definite association of hippocampal LTP with memory was demonstrated [11].

An 'on and off' switch similar to the binary system of computer memory based on phosphorylation and dephosphorylation of CamKII was proposed (review, [12]). Later, the problem was approached by asking "What changes should be induced in neurons to assist the retrieval of memory?" The role of synaptic tag was proposed [13]; however, work is continuing to find the answers. Addition or deletion of new connections between neurons was suggested to increase memory storage capabilities; but at timescales slower than required (review, [14]). A self-propagating prion-like protein was a candidate for memory [15]. Quantum super positions of different arrangements of tubulin molecule within microtubules termed as 'objective reduction' were suggested to be capable of information processing (review, [16]). But experimental results are not available.

According to the synaptic plasticity hypothesis of memory, input events induce changes in neuronal connectivity (synaptic 'weight' changes) and these represent the event itself. The mechanism by which these changes in synaptic strength work back for the retrieval of memory still remains unclear. Investigations have clearly shown that while synaptic plasticity is critical for encoding, hippocampal synaptic plasticity could not be seen involved in memory retrieval or systems level consolidation (review, [17]). Other concerns include lack of evidence to prove its sufficiency (reviews, [18-21]) and possible prevention of further learning due to saturation of synapses with LTP (review, [22]).

Protein synthesis, trafficking and modifications were shown accompanying different stages of memory process. However, the time-scales of acquisition or retrieval of different types of memories could not be correlated to these changes. In addition, instability of the newly formed proteins itself indicates that protein changes may only be a part of the preparatory process. Possible loss of memory by overwriting the established pattern of synaptic connectivity has led to the suggestion for imperative radical modifications of the standard models of memory storage [23]. The present hypothesis was developed by taking into consideration the synaptic function, plasticity, network connections, mechanism for retrieval and optimal timescales required. It was evolved partially from published structural data and partially from its well-fitting nature to explain the features of memory and other brain functions.

How can memory be achieved within the physiological timescales when a cue stimulus makes a search? In a reduced form it requires the following characteristics. *In a network, the specificity of memory depends on the subset of postsynaptic membranes that undergoes changes during retrieval, out of the set of postsynapses where changes took place during learning.* This is the general paradigm upon which the present structure-function hypothesis is made. This work incorporates the proposal that the laws of classical physics, rather than quantum mechanics, are sufficient to explain the physiological process of memory (review, [24]). Based on the above and the structural information from the literature, a working definition of memory is made and expanded using topological and functional arguments. An abstract form of the work is published elsewhere.

The structure-function

From a functional point of view, the net effect of activating a neuron is depolarization of postsynaptic membranes of the synapses at its axonal terminals. For an infinitesimal period of time after the postsynaptic activation, the physical presence of the presynaptic neuron may not be necessary for the excitatory postsynaptic potential (EPSP) already on the postsynaptic membrane. This leads to the next assumption that in a normally functioning synapse if the postsynaptic membrane can be depolarized without stimulating its presynaptic neuron, it gives an equivalent effect of activating the presynaptic neuron. During retrieval, equivalent changes that took place during learning can be made if the corresponding postsynaptic membranes that were activated by learning-stimuli can be depolarized. Therefore, mechanisms for activation of a specific postsynapse without the requirement of the primary stimulus that activated its presynapse can constitute a basic unit for memory. Both synaptic and ECM features that support this function are hypoth-esized and evidence for their presence is shown from the literature.

Characterization of the structure

The following assumptions were made. 1) Learning is a relative process that involves the acquisition of new information in relation to others that are either novel or already learned. These latter cues are often embedded within the item to be learned and may not be prominent as a separate stimuli that are used in conditioning experiments. 2) Retrieval does not occur randomly; rather it is evoked by an internal or external cue stimulus. 3) Correlated activation of at least one pair of synapses is involved in learning.

Let **A** and **C** be the presynaptic terminals from neurons **N1** (cue) and **N2** (item to be memorized) and **B** and **D** be their corresponding postsynaptic terminals. **A** and **C** are co-activated by the cue and the item to be learned respectively. (In the present hypothesis, learning is considered as novel without any relation to past learning). When one tests for memory using a cue (external or internal) that generates an action potential (AP) reaching presynapse **A**, what can induce a cellular event in postsynapse **D** (that otherwise can be achieved only by an AP reaching **C**) in the absence of a stimulus from presynapse **C**? Let the structural answer be **X**. Whenever an AP reaches presynapse **A** (cue) and provided **X** can be hypothesized, a cellular event at the postsynapse **D** can be induced in the absence of a stimulus from the learned item. This satisfies the requirements of a system that has memory capabilities. A structure that can achieve this input-independent depolarization of postsynaptic membrane was then investigated.

B and **D** should be functionally connected for substantiating the timescales of the order of 0.5 seconds for normal conscious processes [25]. A feasible structure of **X** that can support these requirements is derived as follows. Side by side arrangement of multiple postsynaptic membrane segments (PSMSs) forms *shared postsynapse* that synapses with corresponding independent presynaptic terminals (Fig.1A). In other words, they have multiple spine heads on a single spine neck. Similarly, postsynaptic membranes of synapses on selected independent spines can be connected by an ionic mechanism through a common ECM between the postsynaptic membranes named as *exneuron* (Fig.1B). Exneuron is a term coined for this work for the convenience of expressing the potential ECM space between independent synapses that are likely to be functionally connected during learning. The structural characteristics of both shared postsynapses and exneurons endow required functional features for memory.

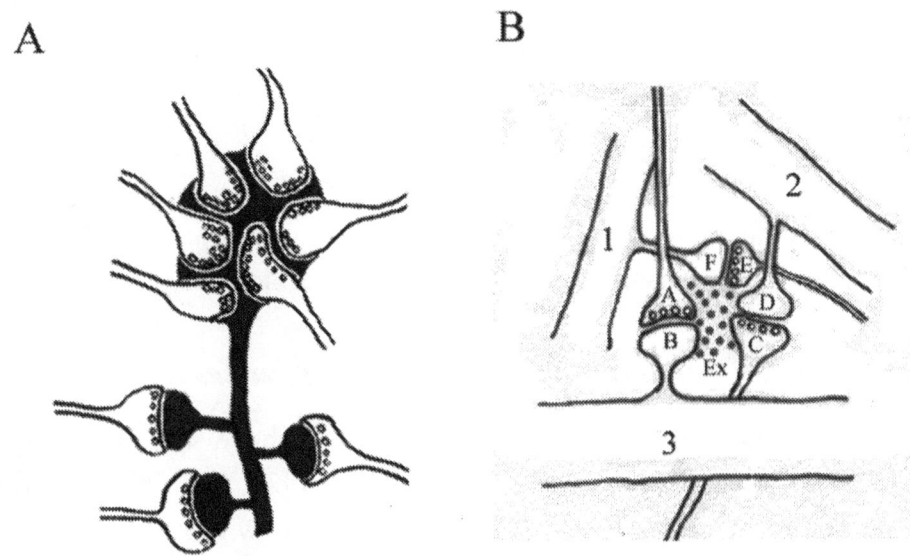

Figure 1. Diagrams of shared postsynapse and exneuron. A) A dendrite showing the presence of a shared postsynapse at its tip with six presynaptic terminals synapsing to it. Note the presence of single spines on the stem of the dendrite with single synapses. B) Synapses on the spines (B, D and F) from three different dendrites (3, 2 and 1) are formed around the ECM space named as exneuron.

Functional semblance

During cue-induced memory retrieval, an AP reaching synapse of **A** can induce a cellular event on PSMS **D** generating an equivalent effect of a primary stimulus reaching **C**. This postsynaptic cellular event that can otherwise be achieved only by an AP reaching **C** is accomplished by depolarizing graded membrane potentials from PSMS **B** to **D** in shared postsynapses or by depolarization or instantaneous activation of channel proteins on PSMS **D** by cue induced ionic changes reaching through the exneurons. Thus, stimulus-independent changes in **D** give rise to cellular illusion of an AP-induced synaptic transmission from **C**. When illusions occur at the multiple shared postsynapses and exneurons in the network of neurons for the memory to be retrieved, they result in "functional semblance" for memory. The suggested mechanism is for excitatory neurotransmission. Projection neurons having different neurotransmitter contents synapse to shared postsynapses and modulate neurotransmission and functional semblance.

Structural evidence from previous studies

Electron microscopic (EM) studies have shown that a single CA3 dendrite has unusual spines with 1 to 16 branches [26, 27]. In this study, large CA3 spine branches were shown to synapse with multiple mossy fiber boutons (presynaptic terminals) indicating the presence of the shared postsynapses. Similar results were also observed at the axonal terminals of the cortico-thalamic projections [28]. Sharing of the same postsynapse was also observed at the excitatory synapses in the rat cerebellum [29].

Thorny excrescences are large clusters of postsynaptic terminals and are seen between the mossy fibers and dendrites of the CA3 pyramidal neurons in the hippocampus [30-32]. Sharing of postsynapses in this region is visible in the EM pictures [33]. In one study, human hippocampal CA3 neurons were shown to have multiple spine heads (4 on average, sometimes more than 10) on a single spine neck [34]. Moreover, thorny excrescences were shown to exhibit spatial changes after memory tests [33]. Thorny excrescences were also studied at the axonal terminals of cortico-thalamic neurons [28]. A similar structural variant where more than one presynaptic terminal synapses with postsynaptic membrane are complex spines present in the cerebral cortex [31, 35]. If shared postsynapse is a universal structural feature that supports memory, then it should be present at least in a simple form in lower level species. This is evident by the presence of more than one presynaptic terminal converging on to one postsynapse in the central nervous system of locusts and aplysia [36, 37].

Suggested cellular changes

The following are suggestions of how changes can be brought about by coincident activation of presynaptic terminals during learning.

1. At the shared postsynapses: Novice PSMSs are likely functionally independent so that depolarizing graded membrane potentials will not reach the neighboring PSMS due to the presence of a surrounding ring of resistant membrane (Fig.2A). During coincident activation of two presynaptic terminals synapsing on the shared postsynapse, functional connection is established between their PSMSs by the following possible changes in the intervening zone (Table 1).

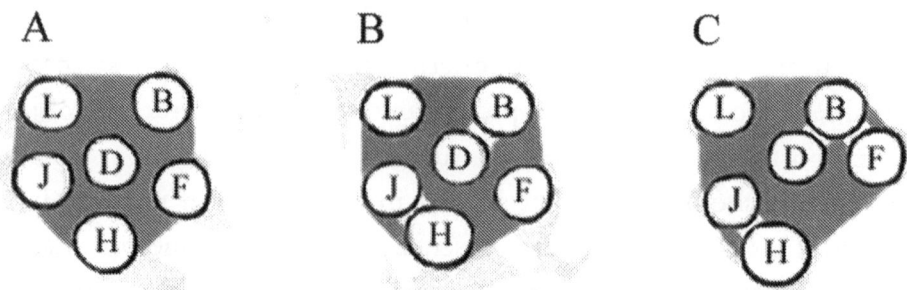

Figure 2. Diagrams of 2-D surface view of a flattened shared postsynapse. A) Six postsynaptic membrane segments (PSMSs) named B, D, F, H, J and L are shown. The intervening membrane is shaded. B) Coincident activation of the corresponding presynaptic terminals of PSMSs B and D induces functional connection between them. Note similar change between J and H. C) Coincident activation of B and F during a separate incident of learning induces functional connection between B and F. This results in the formation of a functional group consisting of PSMSs B, D and F.

a.	Insertion of voltage-activated Na⁺ channels along the connecting resistant membranes between the coincidently activated PSMSs
b.	Lateral displacement of resistant membranes between the PSMSs result in the latter's apposition (Fig.2B)
c.	The intervening ECM changes favor electrical continuity between the PSMSs

a.	Insertion of voltage-activated Na^+ channels along the connecting resistant membranes between the coincidently activated PSMSs
b.	Lateral displacement of resistant membranes between the PSMSs result in the latter's apposition (Fig.2B)
c.	The intervening ECM changes favor electrical continuity between the PSMSs

Table 1 Possible changes at the shared postsynapses during coincident activation of two presynaptic terminals.

Even at low density, voltage-gated Na^+ channels that are reported to be present on spines [38] can boost graded depolarization with less spatial decrement in amplitude [39]. In addition, over short distances, depolarizing graded membrane potentials are transferred more quickly [40]. During a novel learning by a single trial, changes are initiated to bring the co-activated PSMSs closer in all the orders of neurons from both the new material and the cue. However, immediate functional connections are established only in those shared postsynapses where the corresponding PSMSs are already close (Figs.2C, 3B). In shared postsynapses where they are not close, continued repetition of learning induces movement of these PSMSs towards each other until they are functionally connected.

Without any repetition, the initiated changes in shared postsynapses where the PSMSs are farther apart will eventually stop. In learning a related item, one of the PSMSs that took part in previous learning may be co-activated with an unrelated new one on the same shared postsynapse. These newer associations added up with experience result in groups of functionally connected PSMSs on shared postsynapses (Fig.2C). As a result, groups of functionally associated PSMSs that are separated from other groups will be seen. Activation of one PSMS will lead to activation of others within a PSMS-group and may contribute to sensitivity for memory.

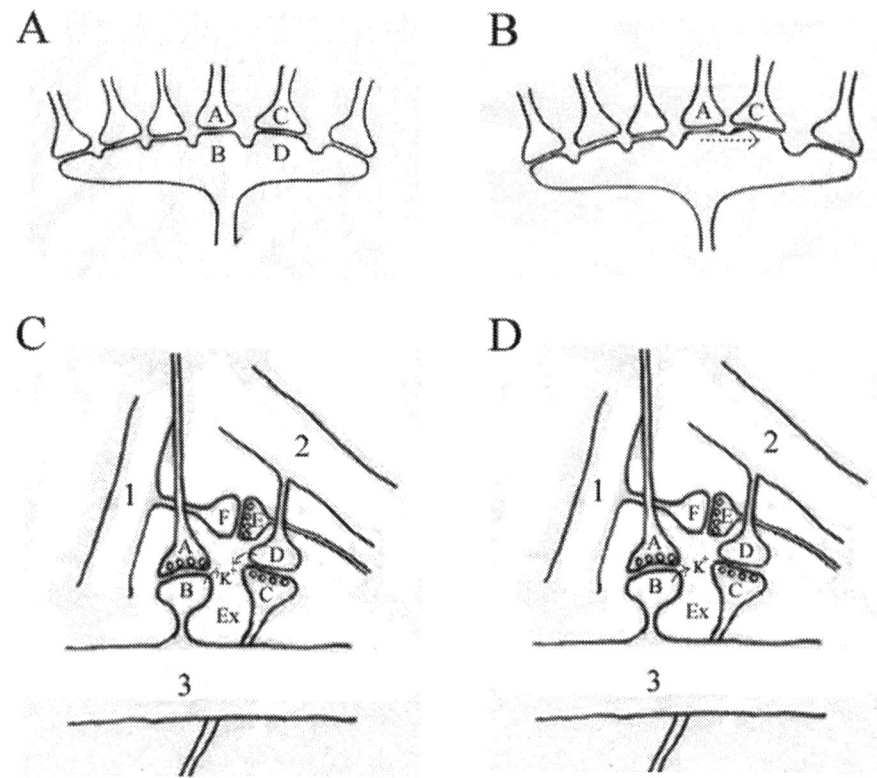

Figure 3. Diagram of suggested changes in shared postsynapses and exneurons. A) Shared postsynapse showing co-activation of presynaptic terminals A and C. B and D are their corresponding post synaptic membrane segments (PSMSs) that will induce changes shown in Fig. 2B. B) AP reaching A induces depolarization of B which spreads to D evoking an equivalent effect of an AP reaching presynaptic terminal C. C) Co-activation of two postsynapses around an exneuron induce ECM changes to improve ion-exchange capability of the exneuron. D) During memory retrieval by the cue, repetitive activity-induced extracellular accumulation of potassium and reduction in calcium ions (not shown) by postsynapse B around the exneuron will induce two types of changes: first, depolarization of the postsynapses in its vicinity; second,

the sequential ionic changes in the exneuron reminiscent of equivalent sequential ECM changes that can otherwise result from depolarizing other postsynapses around the exneuron.

2. At the exneurons: Both theoretical and laboratory studies have shown that changes in extracellular ionic changes brought about by the activation of one neuron can influence another one in its immediate vicinity [41, 42]. During large population spikes in the hippocampus, field effect depolarizations of approximately one-half the population spike amplitude are produced in non-firing pyramidal neurons [43]. Neuronal activity increases extracellular potassium ion concentration [44, 45] and decreases extracellular calcium ion concentration [46, 47]. In addition, reducing the extracellular calcium ion concentration increases membrane excitability [48]. It is likely that co-activation of two postsynapses around an exneuron can induce ECM changes to improve ion-exchange capabilities (Fig. 3C). During memory retrieval, activity-induced extracellular accumulation of potassium ions by one postsynapse around an exneuron can induce depolarization of the postsynapses in its vicinity. Alternatively, sequential ionic changes in the exneuron produced by depolarization of one postsynapse around the exneuron will bring sufficient changes on the membranes of other postsynapses (Fig.3D) that contribute to the functional semblance.

Features of shared postsynapses and exneurons

1. Structural plasticity changes: When two functionally unrelated inputs reach a shared postsynapse during learning, depolarizing their corresponding PSMSs, immediate early changes occur between them (Table 1). In addition, the cellular machinery at the postsynaptic zone of the shared postsynapse will try to bring the co-activated PSMSs closer in order to facilitate sharing of the molecular resources in the common postsynaptic compartment. Synaptic plasticity studies have shown that alterations in spine shape or density are its enduring structural correlate (review, [49]) and have been reported to accompany learning of various tasks. Similarly, learning can induce specific changes in the three dimensional (3-D) space between simultaneously activated PSMSs within the shared postsynapse. The co-activated PSMSs move closer by displacing the intervening synapses, if any, between them (Fig.4). Compared to the single spines that elongate and retract in one dimension, PSMSs of shared postsynapse can reorganize their positions in 3-D and is highly ATP and protein synthesis dependent.

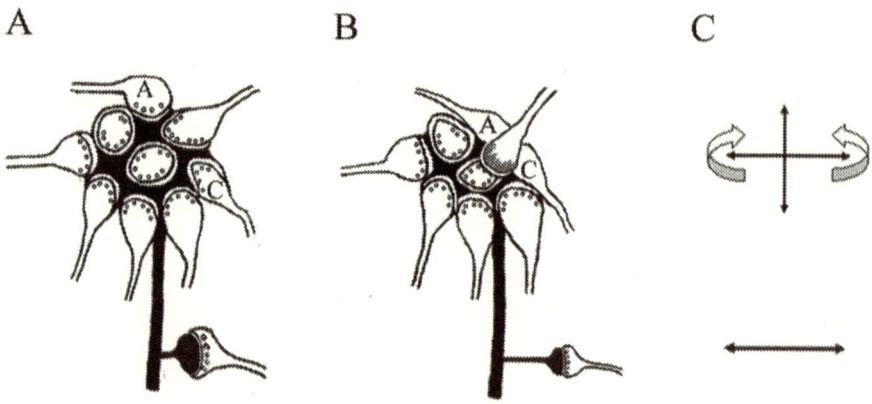

Figure 4. Diagram of change in positions of some of the synapses in a shared postsynapse. A) Cross-sectional view of a dendritic branch terminal with a shared postsynapse at the tip and one single spine on the dendritic stem. The actual number of presynaptic terminals is likely to vary. B) Single spine on the dendritic stem elongates in one direction, whereas shared postsynapse at the tip of the dendrite changes shape in all directions bringing closer the co-active presynaptic terminals 'A' and 'C' without disturbing relations with others. Note that the intervening synapse is pushed towards the viewer in the z plane. In real-

ity, a part of the presynaptic terminal may only get moved while the other part retains the previously established relations. Spaced repetition of learning will be required for an effective change resulting from a completely novel learning. C) Directions of change in single spines and shared postsynapses.

2. Retrieval of memory: After learning, the cue stimulus reaching one presynaptic terminal of either a shared postsynapse (Fig.3B) or an exneuron (Fig.3D) will induce cellular events at the postsynapses of the neurons of the learned item. The cue stimulus evokes memory depending on the net functional semblance from different orders of shared postsynapses determined by factors listed in Table 2. Weight of the changes in shared postsynapses at higher orders is likely different from that of the initial orders.

a.	Initial distance between the synapses of the neurons from the learned item and the cue within the shared postsynapses
b.	Previous related learning events that have already established relations in shared postsynapses
c.	Speed of movement of the PSMSs in shared postsynapses at different orders of neurons due to learning
d.	Total number of the shared postsynapses and exneurons involved at each order of neurons
e.	Total number of orders of neurons involved
f.	Time elapsed after the last repetition of learning
g.	Threshold number of effective shared postsynapses and exneurons required for functional semblance
h.	Number of repetitions of learning at optimal intervals to reach the asymptote for retaining memory for a required period of time
i.	Strength of the cue for evoking functional semblance for a specific memory
j.	Factors specific to the location within the central nervous system
k.	Neurotransmitter types in the terminals reaching shared postsynapses and their release depending on emotional states, for example, dopamine release during motivation.

Table 2 Factors that determine net functional semblance at the shared postsynapses and exneurons from different orders neurons.

3. Ease of learning a related task: In a related learning event, one presynaptic terminal (whose PSMS is already connected with another as a result of previous learning) may become co-activated with different presynaptic terminals within a shared postsynapse. Each new learning event will add new PSMS relations and optimize the spatial positions of the PSMSs within PSMS-groups. This gives a possible cellular explanation for the ease with which new information can be incorporated in the presence of an associate "schema" for fast system consolidation during related learning [50]. Both repetition and experience will eventually result in increased sensitivity to partial cues. Thus, learning an interrelated task requires less energy expenditure due to the preservation of relations between PSMSs from previous learning. This is in agreement with the positron emission tomographic (PET) results of fewer neural activity requirements to process repeated stimuli for memory [51]. Due to the presence of interconnected groups of PSMSs, when a cue stimulus reaches presynaptic terminals during retrieval it may induce nonspecific activation of some of the PSMSs at certain neuronal orders; however, these will fail to achieve functional semblance due to the lack of activation of corresponding PSMSs in the remaining orders of neurons.

4. Disuse reduction in memory: After the first session of new learning, absence of optimally spaced repetitions will likely result in withdrawal of the newly initiated cellular changes (Fig.4). Recent novel learning without any relation to the previously learned tasks and those without optimally spaced repetitions may not result in long-lasting changes. In the case of a learned task, occasional repetition of learning or events of related learning will maintain the structural relations between the functionally related PSMSs. Changes brought about by coincident activation of postsynapses in exneurons are not likely stable compared to those occurring in nearby PSMSs in the shared postsynapses. But immediately after learning they may contribute to functional semblance and may have a role in working memory.

5. Degeneracy at the shared postsynapse: The rules set for defining the structure under the 'structure-function' title need not be strictly followed. That is, firing of **A** need not be necessary for activation of postsynapse **D**. Firing of a third presynaptic terminal in the shared postsynapse or an exneuron may also cause activation of the postsynapse **D**. This means that the cue stimulus that reaches the third presynapse is not the one that was used during learning, but it is one associated with the cue in another learning (cue of the cue). This results from functional connections within a PSMS-group (Fig. 2C) resulting in degeneracy to the memory code at a single shared postsynaptic level and provides sensitivity and flexibility for retrieval of memory using limited features from a related cue.

6. Strength of the cue stimulus and functional semblance: A cue may be a component within the learned task or separate from it. One cue stimulus may activate a large number of sensory neurons reaching a large number of shared postsynapses. For perfect learning and retrieval, a specific cue that reaches a specific set of shared postsynapses is favored. Gene knock out of NR1 subunit of NMDA receptors from CA3 neurons failed to retrieve memory using a partial cue [52] indicating that inputs from cue-specific details are important in achieving functional semblance. In addition, this shows that glutamatergic synapses are major components of the shared postsynapses.

7. Inhibition of protein synthesis causes de-routing of semblance: Repetition of learning is associated with structural changes to optimize spatial relations between the PSMSs on the shared postsynapses. Blocking protein synthesis during repetition of learning alters the relations of synapses on shared postsynapses depending on the requirement of different structural proteins at the active synapses, protein half-lives and presence of precursor forms of proteins. These factors may also affect structural features that isolate unrelated postsynapses. Since the functional semblance depends on maintenance of the related alignment of PSMSs on the shared postsynapses, blocking protein synthesis will result in their structural perturbation de-routing semblance to unrelated synapses leading to loss of specificity of memory.

8. *'Synaptic'* to *'systems-level'* memory: Functional semblance is determined by factors (Table 1) from different orders of neurons. The function of shared postsynapses from synaptic to systems-level can be explained by the following example (Fig.5). Suppose visual quality of an object were associated with its location during a learning protocol using a sufficient number of trials. Once learning is complete, memory for object location can be retrieved using visual object quality as a cue. Stimuli from the cue first reach the infero-temporal cortex, and then propagate to a higher order of connections first in the hippocampus followed by the cortex. Functional semblance at the PSMSs on shared postsynapses and exneurons at different orders of neurons along the original pathway for stimuli from object location leads to memory for locating the object. Hippocampus receives different processed sensory input forming initial orders of shared postsynapses. Since the number of synapses that one neuronal axon can make (10^4) is closer to the number of pyramidal neurons in the hippocampal sub-region (review, [53]), multi-sensory inputs can reach PSMS-groups in large shared postsynapses in the CA3 (as predicted from the size of the excrescences) contributing to sensitivity for memory

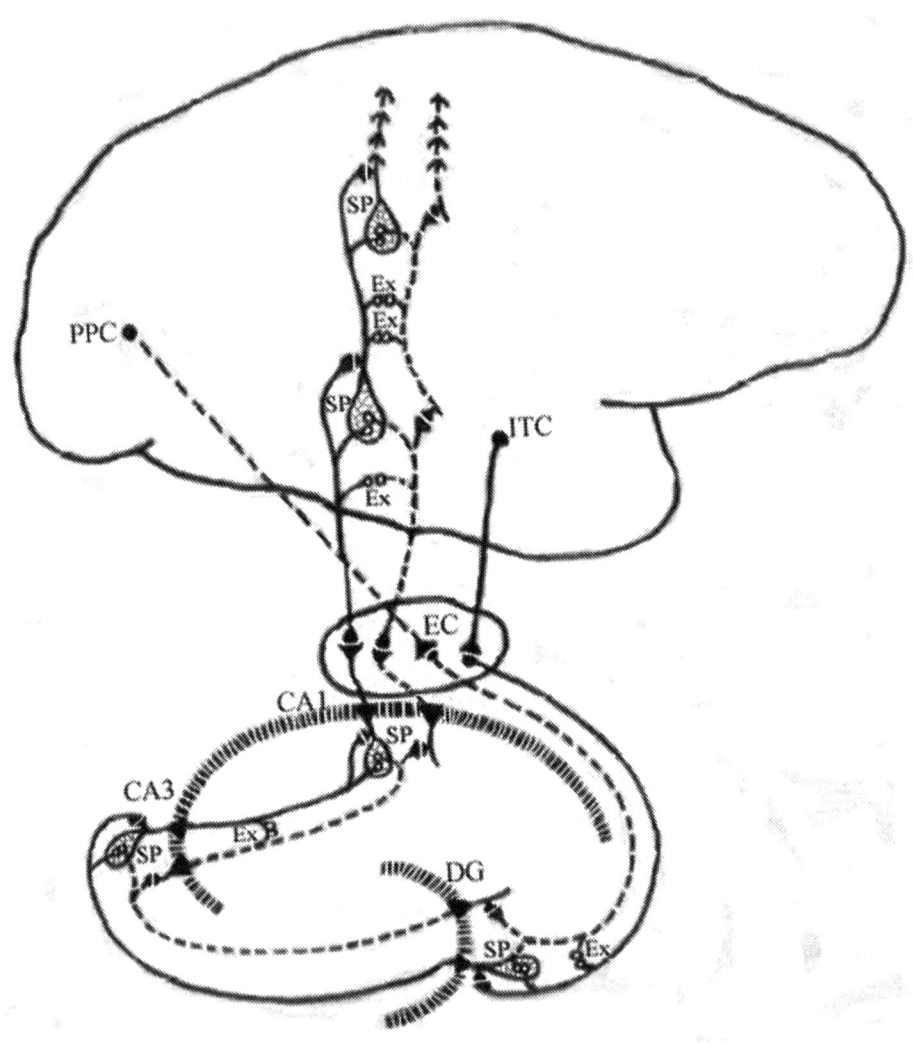

Figure 5. Schematic representation of a learning process through shared postsynapses in hippocampus and cortex. Hippocampus receives putative connections from the infero-temporal cortex (ITC) representing visual object quality (dark lines) and from the posterior parietal cortex (PPC) representing object location (dashed lines). At the time of learning, information from these cortices reaches the entorhinal

cortex (EC) and then to the dentate gyrus (DG), CA3 and CA1 in the hippocampus. Finally, they enter the cortex through EC. Learning will require changes in shared postsynapses (SP) as explained in the text. Note that within each shared postsynapse (SP), postsynapses (due to space constraints only postsynapses are drawn (in circles)) from neurons from the two pathways are in close apposition (drawn in dark color) complete learning. Functional connections between the cue and the learned item also take place through the exneurons (Ex) represented as closely apposed postsynapses (due to space constraints only postsynapses are drawn (in circles)). The final pathways entering the cortex are drawn without any specifications of their connections. During retrieval, stimuli reaching from visual object quality (dark lines) induce functional semblance for the object location (dashed lines) through both the shared postsynapses and exneurons. Connections through the subiculum, recurrent collaterals in the CA3 and several other projections from the medial temporal lobe are not shown.

In the cortex, the number of pyramidal neurons far exceeds those in the hippocampus and the size of shared postsynapses is likely smaller than those at CA3 dendrites. Here, the probability of occurrence of presynaptic terminals of higher orders of neurons from a specific cue and the item to be learned within a single shared postsynapse or around an exneuron will be less and when present, contributes to specificity (similar to that of a single arrangement in permutation). By specializing in one field of expertise, related learning results in frequent stimuli reaching specific cortical areas resulting in more functional connections through the shared postsynapses and exneurons. Exneuron changes may contribute more towards functional semblance in the cortex than hippocampus.

9. Consolidation of memory: During hippocampus-dependent learning, specific inputs reach shared postsynapses at various orders spanning from the hippocampus to the cortex. Optimally spaced repetitions of learning will augment protein synthesis required to bring those distant co-activated synapses within the shared postsynapses closer. Upon completion of spaced repetitions, near-permanent changes occur in shared postsynapses in all the orders of neurons. At this time if the hippocampus is removed, it will leave the alignment of PSMSs in shared postsynapses outside the hippocampus and within the exneurons nearly intact allowing effective functional semblance for memory when a cue stimulus reaches those extra-hippocampal shared postsynapses and exneurons.

10. Hippocampal neurogenesis: Axonal terminals from the hippocampal new granule neurons introduce new connections to the shared postsynapses on the CA3 dendrites. Concurrently, an equal number of neuronal cell deaths are also expected resulting in the loss of presynaptic terminals to CA3 shared postsynapses. Due to this replacement of presynaptic connections to the CA3 shared postsynapses, repetition of learning induces changes in different sets of shared postsynapses. This can have different effects for hippocampus and the cortex. Even if a presynaptic input from the cue stimulus to a shared postsynapse is lost in the hippocampus, the cue stimulus can still reach other postsynapses and exneurons in the hippocampus and cortex and can induce memory. The relationships that their axonal terminals make on shared postsynapses of the CA3 dendrites change depending on the half lives of the granule neurons. Optimal repetition of learning both before and after the replacement of one granule neuron in the pathway can improve specificity of learning through cortical effects. Similarly, shared postsynapses on the dendrites of granule neurons will also bring additional effects. Repetition of learning after the introduction of new neurons in the hippocampal pathway will result in an increase in specificity for memory (similar to the increased specificity of single combination when the number of postsynapses from which to choose is increased as in permutation) by increased functional connections through additional shared postsynapses.

11. Hippocampus and reconsolidation: It was reported that during reconsolidation the memories are labile and are protein synthesis-dependent. Since the last learning, new granule neurons have made new relations at the shared postsynapses of the CA3 dendrites and on their own dendrites. The same old cue stimulus now activates new granule neurons and APs reach through both the old (resulting in semblance for memory) and new granule neurons (resulting in a new set of PSMS relations) in the shared postsynapses of the CA3 and exneurons. Protein synthesis-dependent isolation of these two sets of connections is a likely outcome. If protein synthesis is blocked at this stage, the relations will be de-routed in the shared postsynapses. When retrieval is made after this, the de-routed semblance from the hippocampus having higher sensitivity dominates over the semblance from the cortex resulting in loss of memory.

12. Effect of multiple neurotransmission on the shared postsynapse: Since the shared postsynapse is a large structure, release of one neurotransmitter binding to the receptors on a PSMS can influence the action of others on the shared postsynapse. For example, neurotransmitter dopamine implied in motivation and needed for attention-requiring explicit memory gets released from one of the presynapses, bind to the receptors on the shared postsynapse and can induce changes in the functioning of the whole shared postsynapse.

13. Potential to explain brain oscillations: Working memory task is associated with cortical theta oscillations in humans [54, 55] and in rodents [56]. Along with the high-frequency firing of inhibitory interneurons, coupling between the axons of pyramidal neurons is important for generating or modulating oscillations (review, [57, 58]. Activation of the axons was reported to induce antidromic activation of neighboring axons resulting in somatic spikelet potentials in neurons of hippocampal slices [59]. Even though axonal gap junctions were shown to maintain this coupling function [59, 60], experimental evidence is inconclusive. Can shared postsynapses function as electrical junctions transmitting AP from axon to axon? Reduced ECM volume fraction is a contributory factor for electrical field effects [43] and it is likely to reduce the current required to charge the membrane capacitor and drive the membrane potential to the threshold level. It is also reported that surrounding ECM electrical features may contribute to the generator potentials at sub-thresholds for AP generation [41]. Multiple presynaptic terminal inputs in groups of PSMSs in shared postsynapses may augment this effect and favor axo-axonal AP propagation. This is a possible outcome at least in certain groups of PSMSs and can likely contribute to the neuronal oscillations.

14. Sufficiency and limitations of the hypothesis: In the cerebral cortex where the majority of neurons are pyramidal, considering that the number of shared postsynapses per pyramidal neuron is between the ranges of 10^1 to 10^3, the combinations that lead to functional semblance in a network of nearly 10^{11} neurons appears to be large. One study showed that in the hippocampus, CA3 pyramidal neurons have 41 excrescences [32] per neuron. In humans, some excrescences have more than 10 postsynapses on them [34]. These features highlight possible important functional roles. How can we experimentally prove semblance hypothesis? Application of advanced techniques in an appropriate system will be a major required step.

Shared postsynapses, LTP and memory

Experimental LTP is often induced by high frequency stimulation with an electrode kept over a fiber tract whose fibers synapse to a neuron or groups of neurons. Stimulating the Shaffer collaterals containing fibers from different sensory modalities is equivalent to potentially transmitting signals from many cues and learning items *in vivo*. This multi-fiber stimulation leads to coincident activation of a large number of PSMSs in multiple shared postsynapses and exneurons. The observed LTP of a large number of synapses to a neuron is different from the operating principle of functional semblance. Since a single neuron receives a multitude of synapses, LTP induction protocols by stimulating a large number of fibers not involved in a particular learning is not suitable for explaining semblance hypothesis.

Do shared postsynapses contribute positively or negatively to the LTP? Can it support the essential features of LTP induction [61] that explains synaptic tag? Can it explain the association of LTP and fear conditioning [62]? During LTP induction, the NMDA receptor channels on the membranes between PSMSs of coincident stimuli stay open for long duration resulting in the movement of the molecules of downstream pathways to a common area beneath the PSMSs. In addition, coincident activation of multiple fibers by the stimulating electrode creates large groups of connected PSMSs. These can overlap smaller groups that will be created by a weak stimulus to the synapses of the same neuron at a different location. This allows the weaker stimulus to induce opening of the channels in the entire large groups and induce late-LTP and support the findings that explain the features of synaptic tag proposal [61].

Consequences

Identification and characterization of shared postsynapses and exneurons are the major steps required to prove the hypothesis. One obstacle for visualizing shared postsynapses is the fact that sections used for EM are very thin (5×10^{-8}m) and this prevented reconstruction studies from detecting shared postsynapses of larger sizes ($>5 \times 10^{-6}$m) and non-uniform cross sectional shapes. There were no alternative methods to visualize an isolated shared postsynapse in 3-D with a good resolution. Moreover, an increasing number of relations between synapses result in thinner, tortuous and multi-faceted shared postsynapses that are difficult to identify as a single structural entity. Previous studies using EM focused on normal synaptic features or simple variations of it and were not aimed at searching for this specific structure. However, closely placed synapses and small irregularly shaped dendrites observed using EM [26, 31, 35] may explain some of the characteristic features of this structure in disguise. Golgi staining reaction can reach only up to the 3rd or 4th level of dendritic branching showing single spines on them. Both the expression of fluorescent proteins and injected fluorescent dyes did not help in visualizing shared postsynapses. Either physical or chemical properties of shared postsynapses might have been preventing the access. Chemical manipulation together with improved microscopic techniques may aid in visualizing an isolated shared postsynapse. Standardization of new protocols can be done by examining thorny excrescences that are large clusters of postsynaptic terminals of synapses at the dendrites of CA3 pyramidal neurons.

What post-synaptic cellular event is an equivalent cellular change of an AP reaching its presynaptic terminal? The ideal event, used in the present hypothesis, is depolarization of specific postsynaptic membranes without the requirement of an AP reaching its presynaptic side. Activation of a postsynaptic membrane channel protein that can sense ECM ionic changes, for example Na/K-ATPase, may be important for semblance through exneurons. As we move towards intracellular structural and globular proteins, suitability decreases due to reduced chances of meeting the physiological time scales and non-specific activations. Can semblance form at closely positioned postsynapses on the soma? Since spine necks are shorter or absent at the soma, the large number of PSMSs can form large number of interconnections that might increase sensitivity for memory at the cost of specificity. Improvement in nanotechnology and imaging techniques will allow for testing of this hypothesis.

Discussion

Learning can modify the nervous system enabling the organism to remember an event or a context or an item in association with another. From the biological point of view, necessary cellular changes—physical, chemical or biochemical that serve as building blocks need to take place upon which an enduring record of memories can be built. Since knockout of different genes separately resulted in loss of memory, the assumption made was that memory is beyond the level of individual molecules and an epimolecular mechanism was sought.

Equivalent cellular events of learning can take place during retrieval in the presence of the cue to induce virtual sense of those acquired during learning. Therefore, a basic question was asked. Can the cue stimulus activate those specific cellular events? If the answer is yes, how can it activate synapses at different levels of the network for memory? The present hypothesis is summarized as follows: a cue stimulus evokes an equivalence of the second stimulus (to be memorized) by virtue of the fact that the latter was stimulated along with the cue in the past, repeated enough to bring spatial alignment between synapses in a threshold number of shared postsynapses and exneurons and evokes functional semblance required for long-term memory.

Optimizing energy expenditure is an adaptive ability of cellular systems. In a shared postsynapse, coincident activation of two postsynaptic segments that are located away from each other results in their movement towards each other to share the molecular resources. Thus, from a structural point of view, the highly plastic nature of PSMSs on the shared postsynapse can be seen only as a cellular adaptation mechanism. Overlapping activation of different sets of inputs into a shared postsynapse will result in groups of PSMSs that can respond to input through any one of its presynaptic terminals. From a functional point of view, this enables activation of many postsynapses that were not coincidently activated during learning. Even though this may reduce specificity of semblance at a single shared postsynapse level, it will not affect the memory arising from the cue stimulus that depends on the net semblance from different orders of neurons. Interestingly, this comes with an advantage for discovering or solving a puzzle using interconnected PSMS groups.

According to the present hypothesis, species with an increased number of large shared postsynapses and with efficient exneurons can have increased memory capabilities. Based on the structural details of shared postsynapse, it is conceivable that it is very difficult to enhance memory, which is in agreement with the lack of evidence for manipulations that enhance memory. On the other hand, memory can be reduced by manipulating proteins that control both the structure and function of synapses as reported previously. Since shared postsynapses function similarly to the other synapses, semblance will be impaired as a result of mutations or knockouts of proteins in the perisynaptic structures. From the developmental point of view, shared postsynapse may result from the disproportionately shorter dendrites that fail to hold all the spines as singlets. Formation of an optimal shared postsynapse might be occurring favorably in humans compared to other species.

The background sensory inputs that an individual receives initiate functional semblance at various shared postsynapses resulting in perception of the external environment without evoking any specific memory. The same mechanisms of cue-induced functional semblance for memory occurring in dysfunctional shared postsynapses may explain delusions and hallucinations due to disease. At an advanced level, semblance can explain the ability to find alternate answers and even formulate a rare possible answer based on synaptic relations on the shared postsynapses and exneurons in the network acquired through previous experiences. The paradigms of an organism may depend on the way learning was done using the cues and this, in turn, depends on the way synapses are made to organize from the very beginning of life. Therefore, these paradigms can decide the outcome of the exposure to new cues and may explain behavior.

Dipoles associated synaptic currents in shared postsynapses are likely to be higher by the order of 10^2 and can be detected at large distances contributing to the surface recorded electroencephalogram (EEG). If lateral AP propagation takes place through shared postsynapse inducing oscillating brain waves as explained previously, the dipoles associated with them will be comparatively larger further contributing to EEG.

Functional semblance at the shared postsynapses is considered complementary to the normal AP generation, propagation and synaptic transmission. In summary, coincident synaptic activations within shared postsynapses and exneurons lead to structural alterations that determine memory capabilities. There is no storage either in the form of potential energy or as modified molecules; memory is a consequence of sharing of structural and functional resources by multiple PSMSs that are coincidently activated and ionic effects at the exneurons. The incorporation of synaptic function, structural plasticity and ability to explain most of the neuronal functions with shared postsynapses and exneurons, makes them candidates for further examination. Semblance hypothesis should be regarded as unproved until it is checked against more exact results.

References

[1] S. Ramón y Cajal, Histologie du Système Nerveux de L'homme et des Vertébrés (Maloine, Paris), Vol.2., (1911).

[2] D. O. Hebb, *Organization of Behavior* (Wiley, New York), (1949).

[3] T. V. Bliss and T. Lomo, Long-lasting potentiation of synaptic transmission in the dentate area of the anaesthetized rabbit following stimulation of the perforant path, *J Physiol* **232** (1973), pp. 331-356.

[4] T. V. Bliss and G. L. Collingridge, A synaptic model of memory: long-term potentiation in the hippocampus, *Nature* **361** (1993), pp. 31-39.

[5] H. Markram, J. Lubke, M. Frotscher and B. Sakmann, Regulation of synaptic efficacy by coincidence of postsynaptic APs and EPSPs, *Science* **275** (1997), pp. 213-215.

[6] D. A. Hoffman, R. Sprengel and B. Sakmann, Molecular dissection of hippocampal theta-burst pairing potentiation, *Proc Natl Acad Sci U S A* **99** (2002), pp. 7740-7745.

[7] D. S. Wei, Y. A. Mei, A. Bagal, J. P. Kao, S. M. Thompson and C. M. Tang, Compartmentalized and binary behavior of terminal dendrites in hippocampal pyramidal neurons, *Science* **293** (2001), pp. 2272-2275.

[8] P. Poirazi and B. W. Mel, Impact of active dendrites and structural plasticity on the memory capacity of neural tissue, *Neuron* **29** (2001), pp. 779-796.

[9] J. Z. Tsien, P. T. Huerta and S. Tonegawa, The essential role of hippocampal CA1 NMDA receptor-dependent synaptic plasticity in spatial memory, *Cell* **87** (1996), pp. 1327-1338.

[10] M. Mayford, M. E. Bach, Y. Y. Huang, L. Wang, R. D. Hawkins and E. R. Kandel, Control of memory formation through regulated expression of a CaMKII transgene, *Science* **274** (1996), pp. 1678-1683.

[11] J. R. Whitlock, A. J. Heynen, M. G. Shuler and M. F. Bear, Learning induces long-term potentiation in the hippocampus, *Science* **313** (2006), pp. 1093-1097.

[12] J. Lisman, The CaM kinase II hypothesis for the storage of synaptic memory, *Trends Neurosci* **17** (1994), pp. 406-412.

[13] U. Frey and R. G. Morris, Synaptic tagging and long-term potentiation, *Nature* **385** (1997), pp. 533-536.

[14] D. B. Chklovskii, B. W. Mel and K. Svoboda, Cortical rewiring and information storage, *Nature* **431** (2004), pp. 782-788.

[15] N. Nishida, S. Katamine, K. Shigematsu, et al., Prion protein is necessary for latent learning and long-term memory retention, *Cell Mol Neurobiol* **17** (1997), pp. 537-545.

[16] R. Penrose, Where is there scope for a non-computational physics? in *From Brain to Consciousness*. The Penguin press, (1998), pp. 167-179.

[17] R. G. Morris, E. I. Moser, G. Riedel, et al., Elements of a neurobiological theory of the hippocampus: the role of activity-dependent synaptic plasticity in memory, *Philos Trans R Soc Lond B Biol Sci* **358** (2003), pp. 773-786.

[18] S. J. Martin, P. D. Grimwood and R. G. Morris, Synaptic plasticity and memory: an evaluation of the hypothesis, *Annu Rev Neurosci* **23** (2000), pp. 649-711.

[19] H. S. Seung, Half a century of Hebb, *Nat Neurosci* **3 Suppl** (2000), p. 1166.

[20] K. J. Jeffery, LTP and spatial learning—where to next?, *Hippocampus* **7** (1997), pp. 95-110.

[21] S. J. Martin and R. G. Morris, New life in an old idea: the synaptic plasticity and memory hypothesis revisited, *Hippocampus* **12** (2002), pp. 609-636.

[22] T. V. Bliss, The physiological basis of memory in *From Brain to Consciousness*. The Penguin press, (1998), p. 89.

[23] S. Fusi and L. F. Abbott, Limits on the memory storage capacity of bounded synapses, *Nat Neurosci* **10** (2007), pp. 485-493.

[24] C. Koch and K. Hepp, Quantum mechanics in the brain, *Nature* **440** (2006), p. 611.

[25] B. Libet, C. A. Gleason, E. W. Wright and D. K. Pearl, Time of conscious intention to act in relation to onset of cerebral activity (readiness-potential). The unconscious initiation of a freely voluntary act, *Brain* **106 (Pt 3)** (1983), pp. 623-642.

[26] M. E. Chicurel and K. M. Harris, Three-dimensional analysis of the structure and composition of CA3 branched dendritic spines and their synaptic relationships with mossy fiber boutons in the rat hippocampus, *J Comp Neurol* **325** (1992), pp. 169-182.

[27] D. G. Amaral and J. A. Dent, Development of the mossy fibers of the dentate gyrus: I. A light and electron microscopic study of the mossy fibers and their expansions, *J Comp Neurol* **195** (1981), pp. 51-86.

[28] P. V. Hoogland, F. G. Wouterlood, E. Welker and H. Van der Loos, Ultrastructure of giant and small thalamic terminals of cortical origin: a study of the projections from the barrel cortex in mice using Phaseolus vulgaris leuco-agglutinin (PHA-L), *Exp Brain Res* **87** (1991), pp. 159-172.

[29] M. Masugi-Tokita, E. Tarusawa, M. Watanabe, E. Molnar, K. Fujimoto and R. Shigemoto, Number and density of AMPA receptors in individual synapses in the rat cerebellum as revealed by SDS-digested freeze-fracture replica labeling, *J Neurosci* **27** (2007), pp. 2135-2144.

[30] M. Frotscher, L. Seress, W. K. Schwerdtfeger and E. Buhl, The mossy cells of the fascia dentata: a comparative study of their fine structure and synaptic connections in rodents and primates, *J Comp Neurol* **312** (1991), pp. 145-163.

[31] R. Murakawa and T. Kosaka, Structural features of mossy cells in the hamster dentate gyrus, with special reference to somatic thorny excrescences, *J Comp Neurol* **429** (2001), pp. 113-126.

[32] R. B. Gonzales, C. J. DeLeon Galvan, Y. M. Rangel and B. J. Claiborne, Distribution of thorny excrescences on CA3 pyramidal neurons in the rat hippocampus, *J Comp Neurol* **430** (2001), pp. 357-368.

[33] M. G. Stewart, H. A. Davies, C. Sandi, et al., Stress suppresses and learning induces plasticity in CA3 of rat hippocampus: a three-dimensional ultrastructural study of thorny excrescences and their postsynaptic densities, *Neuroscience* **131** (2005), pp. 43-54.

[34] M. Lauer and D. Senitz, Dendritic excrescences seem to characterize hippocampal CA3 pyramidal neurons in humans, *J Neural Transm* **113** (2006), pp. 1469-1475.

[35] R. H. Laatsch and W. M. Cowan, Electron microscopic studies of the dentate gyrus of the rat. I. Normal structure with special reference to synaptic organization, *J Comp Neurol* **128** (1966), pp. 359-395.

[36] A. H. Watson and F. W. Schurmann, Synaptic structure, distribution, and circuitry in the central nervous system of the locust and related insects, *Microsc Res Tech* **56** (2002), pp. 210-226.

[37] J. P. Tremblay, M. Colonnier and H. McLennan, An electron microscope study of synaptic contacts in the abdominal ganglion of Aplysia californica, *J Comp Neurol* **188** (1979), pp. 367-389.

[38] R. Araya, V. Nikolenko, K. B. Eisenthal and R. Yuste, Sodium channels amplify spine potentials, *Proc Natl Acad Sci U S A* (2007).

[39] G. C. Taylor, J. A. Coles and J. C. Eilbeck, Conditions under which Na+ channels can boost conduction of small graded potentials, *J Theor Biol* **172** (1995), pp. 379-386.

[40] S. R. Shaw, Anatomy and physiology of identified non-spiking cells in the photoreceptor-lamina complex of the compound eye of insects, especially Diptera. In: *Neurons without impulses* (Roberts, A and BUsh B.M.H., eds) Cambridge University Press, (1981), pp. 61-116.

[41] J. D. Kocsis, J. A. Ruiz and K. L. Cummins, Modulation of axonal excitability mediated by surround electric activity: an intra-axonal study, *Exp Brain Res* **47** (1982), pp. 151-153.

[42] S. D. Erulkar and F. F. Weight, Ionic environment and the modulation of transmitter release., *Trends Neurosci* **2** (1979), pp. 298-301.

[43] R. W. Snow and F. E. Dudek, Evidence for neuronal interactions by electrical field effects in the CA3 and dentate regions of rat hippocampal slices, *Brain Res* **367** (1986), pp. 292-295.

[44] S. D. Erulkar and F. F. Weight, Extracellular potassium and transmitter release at the giant synapse of squid, *J Physiol* **266** (1977), pp. 209-218.

[45] R. A. Nicoll, Dorsal root potentials and changes in extracellular potassium in the spinal cord of the frog, *J Physiol* **290** (1979), pp. 113-127.

[46] M. Galvan, G. T. Bruggencate and R. Senekowitsch, The effects of neuronal stimulation and ouabain upon extracellular K+ and Ca2+ levels in rat isolated sympathetic ganglia, *Brain Res* **160** (1979), pp. 544-548.

[47] C. Nicholson, G. T. Bruggencate, R. Steinberg and H. Stockle, Calcium modulation in brain extracellular microenvironment demonstrated with ion-selective micropipette, *Proc Natl Acad Sci U S A* **74** (1977), pp. 1287-1290.

[48] A. M. Shanes, Electrochemical aspects of physiological and pharmacological action in excitable cells. II. The action potential and excitation, *Pharmacol Rev* **10** (1958), pp. 165-273.

[49] J. E. Lisman and K. M. Harris, Quantal analysis and synaptic anatomy—integrating two views of hippocampal plasticity, *Trends Neurosci* **16** (1993), pp. 141-147.

[50] D. Tse, R. F. Langston, M. Kakeyama, et al., Schemas and memory consolidation, *Science* **316** (2007), pp. 76-82.

[51] L. R. Squire, J. G. Ojemann, F. M. Miezin, S. E. Petersen, T. O. Videen and M. E. Raichle, Activation of the hippocampus in normal humans: a functional anatomical study of memory, *Proc Natl Acad Sci U S A* **89** (1992), pp. 1837-1841.

[52] K. Nakazawa, M. C. Quirk, R. A. Chitwood, et al., Requirement for hippocampal CA3 NMDA receptors in associative memory recall, *Science* **297** (2002), pp. 211-218.

[53] E. T. Rolls and A. Treves, *Neural Networks and Brain Function* (Oxford University Press, 1998).

[54] C. D. Tesche and J. Karhu, Theta oscillations index human hippocampal activation during a working memory task, *Proc Natl Acad Sci U S A* **97** (2000), pp. 919-924.

[55] O. Jensen and C. D. Tesche, Frontal theta activity in humans increases with memory load in a working memory task, *Eur J Neurosci* **15** (2002), pp. 1395-1399.

[56] J. Winson, Loss of hippocampal theta rhythm results in spatial memory deficit in the rat, *Science* **201** (1978), pp. 160-163.

[57] R. D. Traub, A. Bibbig, F. E. LeBeau, E. H. Buhl and M. A. Whittington, Cellular mechanisms of neuronal population oscillations in the hippocampus in vitro, *Annu Rev Neurosci* **27** (2004), pp. 247-278.

[58] B. A. MacVicar and F. E. Dudek, Electrotonic coupling between granule cells of rat dentate gyrus: physiological and anatomical evidence, *J Neurophysiol* **47** (1982), pp. 579-592.

[59] D. Schmitz, S. Schuchmann, A. Fisahn, et al., Axo-axonal coupling. a novel mechanism for ultrafast neuronal communication, *Neuron* **31** (2001), pp. 831-840.

[60] A. Draguhn, R. D. Traub, D. Schmitz and J. G. Jefferys, Electrical coupling underlies high-frequency oscillations in the hippocampus in vitro, *Nature* **394** (1998), pp. 189-192.

[61] K. C. Martin and K. S. Kosik, Synaptic tagging—who's it?, *Nat Rev Neurosci* **3** (2002), pp. 813-820.

[62] M. T. Rogan, U. V. Staubli and J. E. LeDoux, Fear conditioning induces associative long-term potentiation in the amygdala, *Nature* **390** (1997), pp. 604-607.

978-0-595-47002-0

0-595-47002-5

www.ingramcontent.com/pod-product-compliance
Lightning Source LLC
Chambersburg PA
CBHW050336290526
45785CB00006B/2515